the Highland Cow

Colin Baxter Photography, Grantown-on-Spey, Scotland

the Highland Cow

For many people, encountering Highland cattle is an event. A host of other breeds may also please the modern farmer and consumer for their qualities of meat and milk, growth and temperament. But few others have the sheer charisma to make passers-by stop in their tracks, or tempt them to pose by a fence to be photographed near bull, cow or calf. Highland cattle have pulling power, with a presence that somehow seems much greater than their actual size. After all, they're fairly small. This is both by the standards of contemporary breeds of domestic cattle and according to writers over the centuries who have pondered why crofters and farmers in the west of Scotland didn't develop a bigger animal.

The answer is that the Highland breed is ideally suited to conditions at the ocean rim and in the craggy uplands of its native land. It can cope with poor pastures that are lashed by Atlantic squalls for weeks on end.

It can plodge the fringes of peatbogs unscathed, pick its way across a mountain pass, swim a loch and calve in snow if needs be. And when the rain has stopped and the gale calmed to the more usual breeze that ruffles the west Highland grasses, the animal can look drop-dead gorgeous, in a windswept sort of way.

Tousled hair backlit by summer sun, a faint steam of breath rising from glistening nostrils, an elegant sweep of horns arcing out from the broad head: an adult Highlander in its element is a bonny beast indeed. Add a mop-topped calf or two to the group, and few could fail to be attracted.

But this is also the animal that was a mainstay of the economy in the Scottish Highlands and islands for many centuries. At a local level, the house cow could give milk, cheese, butter, hair for weaving and (eventually) hide and leather for a host of uses.

Regionally, the herds gathered in summer for driving south supported many small markets and trades, including that of the blacksmiths who made shoes for cattle who walked new roads from the 18th century onwards. Nationally, the big 'trysts', where Highland cattle by the bellowing thousand were bought by southern dealers, were among the most colourful annual gatherings in Scotland during the last three centuries.

The scale of the droves that brought Highland cattle from the islands and the northlands down, over weeks of journeying, to the markets at the fringe of the Central Belt, not far from Glasgow and Edinburgh, is now hard to comprehend. Like a host of side streams that converged to the main river, as one writer put it a long time ago, the flow of cattle, and their attendant drovers and dogs, moved across glen and hill to reach places such as Crieff, Falkirk and Stirling.

Those days are gone, but the popularity of Highland cattle is on the upturn again, after a period when the breed's fortunes were on the wane. Farmers in parts of the world as far removed from the Scottish Highlands as northern Canada, Australia and New Zealand now appreciate the qualities of hardiness, calm nature, good maternal care and lean meat for which the breed is renowned.

And in Scotland, there are fine herds – or 'folds' – of Highland cattle in many places, with an increasing number of the old black type, once the norm for the breed but gradually ousted by redheads in the 19th century. With well-supported breed societies in many countries and a solid fan base of knowledgeable enthusiasts, the Highland star is in the ascendant again.

So if you see a group of Highland cattle by a roadside field and are tempted to pause and have a closer look, appreciate the moment. You'll be meeting an icon – an animal as redolent of Scotland's western fringe and glens as the tang of seaweed and the aroma of peat-tinged whisky.

Highland cattle formed an important part of upland and island Scotland's past. They are also an attractive element of present-day organic agriculture. And for the breed, in its Highland home or far abroad, the future now seems as bright as those characterful coats.

The Highland cow is one of the oldest registered breeds of cattle in the world. The first 'herd book' was established in 1884.

Spring and early summer used to be the times when huge numbers of Highland cattle were rounded up in the Highlands and Islands to begin their long journey to market. The men who walked and rode, ate and slept with the herds over many weeks that followed were called 'drovers'. The routes they followed with the cattle to reach the markets are still known as 'drove roads'. In the 1700s, the main destination for the drovers was the town of Crieff. Later, thanks to the new rail network, it was Falkirk that became the principal 'tryst', or market, for the drovers. Sitting at the southern fringe of the main Scottish mountains and with reasonable access south to the Lowlands, both Crieff and Falkirk have ready access to both the Highlands and to several Scottish cities. At the peak of the trade in 1850, some 150,000 cattle were driven and sold at Crieff and Falkirk.

There are several old names for Highland cattle. Early descriptions of the breed often use the term 'black cattle', referring to the main coat colour of the small animals that were typical in the west until the 19th century. Others use the term 'Highlander'. 'Kyloe' was a common term for Highland cattle from the west, and is still used by some people today. The name is almost certainly based on the word 'kyle', which comes from Gaelic and means a narrow stretch of water between two pieces of land. Such kyles played a big part in the life of Highland cattle from places such as Skye, the Outer Hebrides and Sutherland in the past. Thousands of beasts would have to swim across such straits as part of their journey to markets far to the south.

Her Majesty Queen Elizabeth II has a herd – or 'fold' – of Highland cattle on her Balmoral Estate.

With its rakish hair and long eyelashes, which provide good protection against flies in summer, a Highland cow has obvious charm.

What makes the horns of Highland cattle stand out – literally – is the way they curve from each animal's head. A bull's horns, in contrast to a cow's, usually turn downwards. Not only do they look impressive, but they are part of what makes an individual unique. Even from a fair distance or seen in silhouette, this allows someone who is familiar with a herd to tell different animals apart. From a cow's point of view, a good pair of upturned horns can be useful protection against would-be calf attackers, such as wolves. This isn't an issue in the breed's native land any more, where the last wolf was killed several hundred years ago. But in northern Canada – one of the many places where Highland cattle have been introduced – some breeders say that horns can still come in handy in this way.

Look at a group of Highland cattle, and it's a fair bet that you'll see several different colours of coat. Ginger-gold and red-brown are common colours these days, but there could also be animals tending to blonde and a few that are black. A couple of centuries ago, the colour scheme in the glens and islands was very different. Black was the top Highland cattle tint by far, especially in the west.

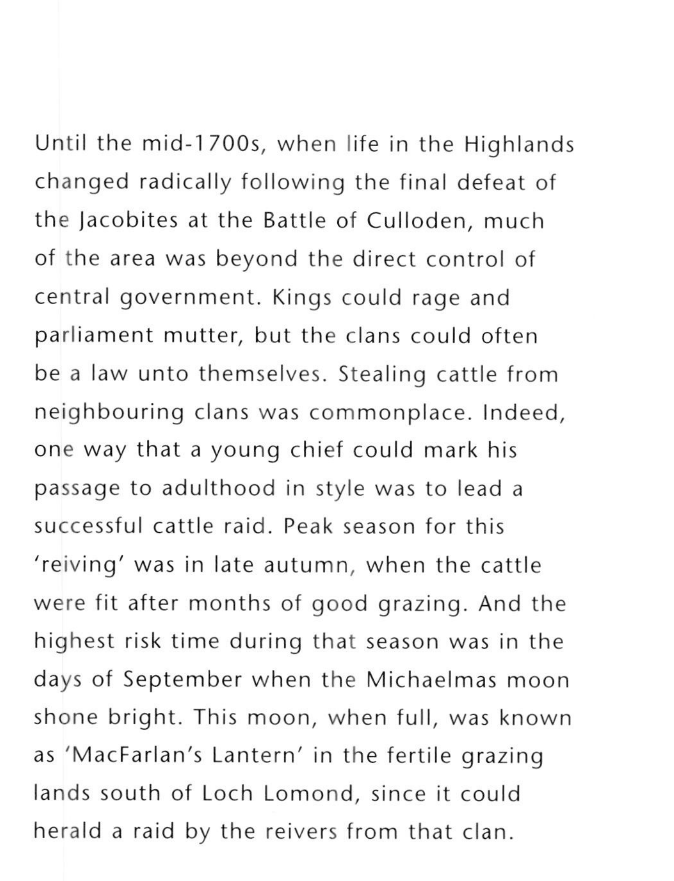

Until the mid-1700s, when life in the Highlands changed radically following the final defeat of the Jacobites at the Battle of Culloden, much of the area was beyond the direct control of central government. Kings could rage and parliament mutter, but the clans could often be a law unto themselves. Stealing cattle from neighbouring clans was commonplace. Indeed, one way that a young chief could mark his passage to adulthood in style was to lead a successful cattle raid. Peak season for this 'reiving' was in late autumn, when the cattle were fit after months of good grazing. And the highest risk time during that season was in the days of September when the Michaelmas moon shone bright. This moon, when full, was known as 'MacFarlan's Lantern' in the fertile grazing lands south of Loch Lomond, since it could herald a raid by the reivers from that clan.

There's more to a Highland cow's hair than wild style. In practical terms, the coat is superb for weathering a climate where summers can be wet and winters cold. The hair grows in a double layer, with a soft undercoat and a long overcoat. This outer coat (which can often be 30 cm long) is naturally wavy and has oils that give it sheen to resist rain. By nature, Highlanders are fairly placid animals and have a reputation as good and attentive mothers who rarely abandon their calves. So if her calf is threatened, a Highland cow will be staunch in defending it. Her usual laid-back approach to life can be cast aside in an instant as she moves quickly to use her horns, or kick out at any person or animal unwise enough to get between her and her offspring.

With its double coat of insulating hair, the hearty Highland cow can calve in snow, and survive outside in temperatures as low as –20° C.

First published in Great Britain in 2005 by
Colin Baxter Photography Ltd.,
Grantown-on-Spey
PH26 3NA, Scotland
www.colinbaxter.co.uk
Reprinted 2011
Text by Kenny Taylor
A CIP Catalogue record for this book is available from the British Library.
ISBN 978-1-84107-283-8 Printed in China